Bruno Bartoletti

Gubbio: Symbols Engraved by Crusaders Returning from the Crusades

The Crusades, from the year 1000 until the mid-thirteenth century, stained the Mediterranean with blood, unleashing power struggles under the banner of Holy Wars, for the redemption of something unknown, but beyond our thoughts, the animosity of the fighters for the holy war led those who returned on their own legs to adorn all the religious structures of their city with glyphs.

They seemed to want to perpetuate the memory of the warrior's effort in the Holy Land.

We don't know if they were clerics, foot soldiers, or knights, but there is evidence throughout our peninsula, especially in the south.

In fact, Danny Vitale writes, "In this context, the multiple symbols on the architectural elements, both internal and external, of the Church of San Giovanni al Sepolcro, a sacred place located in the heart of Brindisi, or one of the forced passages to reach Jerusalem, should be inserted.

The list includes Latin crosses, Lorraine crosses, the triple enclosure, a circle, a Norman ship, and other symbols that are difficult to identify."

Even in Gubbio, we have traces of these engravings in sacred places.

The first one we illustrate is located on the left side of the entrance portal of the church of San Pietro, dating back to at least the mid-11th century.

The building initially belonged to the Benedictine Cassinese, who were replaced by the Olivetan Benedictines in 1519.

The engravings are two and worn by the centuries, but still visible even though, like most of the glyphs we will see, they are difficult to identify.

The one closest to the portal, as seen in the photo, would seem to be a monogram, but it is only a hypothesis.

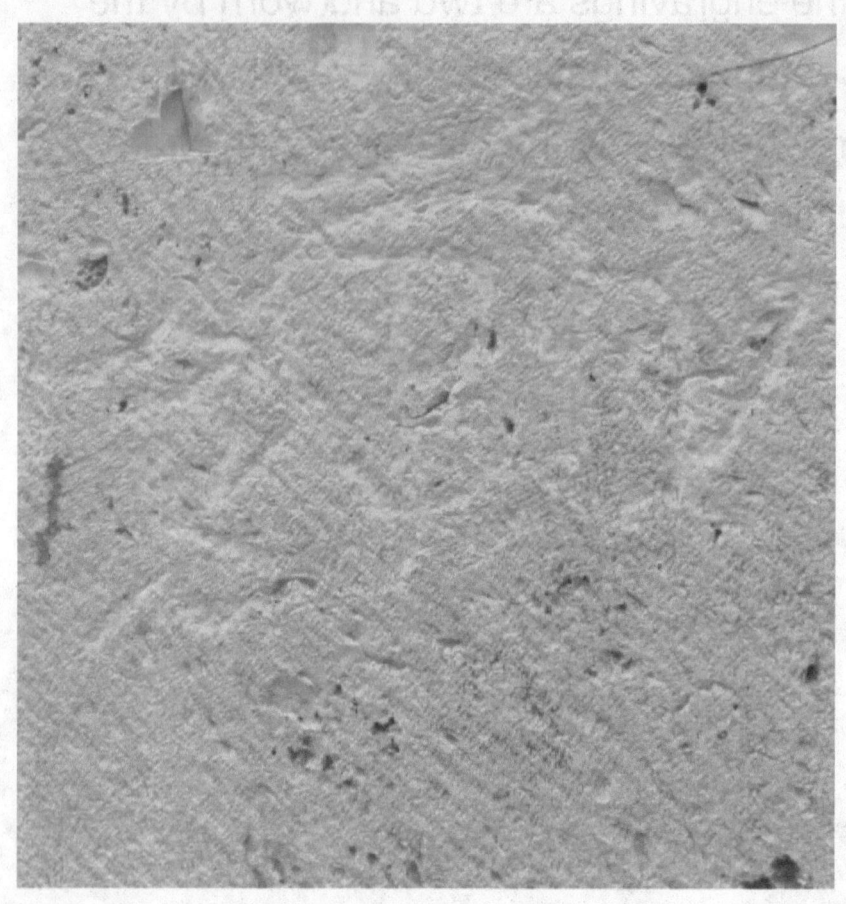

The second, furthest from the portal, seems to contain traces of red pigment and, when observed closely, appears to be a Christogram.

Christograms are combinations of letters
from the Greek or Latin alphabet that form an
abbreviation of the name of Jesus.

They are traditionally used as Christian symbols in the decoration of buildings, furnishings, and vestments.

Some Christograms originated as simple abbreviations or acronyms, although they later became monograms, that is, unified graphic symbols.

Others, like the well-known Chi Rho, were conceived from the beginning as monograms.

Of course, this is also just a hypothesis with no basis because unfortunately, the stone is too worn by time.

Perhaps the most plausible hypothesis is that they were monograms, meant to represent the lineage from which the Crusader Knight came, who had returned home from the Holy Land.

In this way, he connected his war effort with the stone of a sacred place, allowing others to take notice of his gesture made for Christianity.

Two other engravings are found on the side of the church of San Francesco, facing Mount Ingino, where there is another portal, long closed, perhaps an ancient additional entrance.

The church, built in the second half of the 13th century, near the fondaco of the Spadalonga family, which would have welcomed Saint Francis of Assisi after leaving his paternal home, was already officiated in 1256.

Even these engravings, worn by time, are incomprehensible.

However, given the importance of this church, they may have been made by Pilgrims during a journey to some sanctuary or upon returning from Rome during a Holy Year.

Therefore, they could be Christograms, as seen in the image, the symbol of a Cross can be distinguished in the first of the engraved glyphs.

The second is absolutely uninterpretable.

A different situation is that of the glyphs engraved in the Cathedral.

On the left and right sides of the entrance portal, there are engravings recognizable enough as two crosses.

The construction of the current building began based on the design by Giovanni da Gubbio around 1194 on the area granted by Bishop Bentivoglio.

The church was completed in its essential
forms in 1229.

This symbol, engraved on the left side of the
portal, seems, as seen in the photo, to be a
Cross inscribed in a circle, the latter as if to
"empower" the mystical and religious value
of the Cross.

This practice was common among the Knights of the Temple of Jerusalem and had its maximum expression in the Patented Cross, a Cross whose arms spread out towards the ends.

Its name may derive from the Latin etymology "patentem," the past participle of the verb "pateo."

This Cross, in the period following the Fourth Crusade, was often engraved inscribed in a Circle.

On the right side of the portal is another
Cross, this time not inscribed in a Circle but
still "empowered" by transverse lines at the
end of each arm, undoubtedly recalling the
red Cross that the Crusaders wore on their
mantle.

The Templar Cross, in red, was a symbol of the martyrdom of Christ for the redemption of humanity.

It was clearly distinguished from the Maltese Cross of the Hospitaller Knights, whose eight bifurcated ends represented the beatitudes.

Below the Crosses, both to the left and right of the portal, cursive characters or at least that's how they seem, are engraved but absolutely not interpretable, as you can see from the photos.

Climbing towards Mount Ingino, on the left side of the Cathedral, other engravings can be seen next to what once could have been another entrance or window.

They are much higher than those on the main entrance portal but much more intelligible and interpretable, as seen in the images.

On the engraving on the left, in the lower part of the glyph, an anchor can be distinguished.

It was indeed necessary to embark on ships to reach the Holy Land and fight to liberate the Holy Sepulcher.

However, the sea was not always friendly to the Crusaders, and many times, bad conditions due to unexpected weather phenomena had sunk the ships of the Crusader Knights.

This is exactly what some scholars from the American Society of the University of Haifa in Israel discovered in 2017 when they found the remains of a Crusader ship sunk at sea off Acre, a port not far from Jerusalem, in Israel.

A team of underwater archaeologists set out to explore the remains of the sunken ship.

Only small parts of the hull, keel, and planking remained, and the researchers also found ceramics from Cyprus, Syria, and southern Italy, iron nails, and artifacts such as anchors.

Perhaps this glyph was engraved on the walls of the Cathedral of Gubbio by a fellow citizen who had gone to fight in the Holy Land and survived a shipwreck?

Unfortunately, the upper part of the engraving is effectively illegible, so no hypotheses can be made about it.

On the right side, a bird is clearly visible, with feathers represented by many parallel engravings, and the head with the beak pointing towards the East.

In the Middle Ages, the Templars used the symbol of the Phoenix or the Pelican to symbolize Christ.

The Phoenix, or Arabian Phoenix, is born as a very colorful bird with red feathers on the body, a golden neck, and azure in the tail, as well as in one of the two feathers adorning the head.

It had long legs and a tapered beak, a silhouette very similar to that of the heron, although the Romans associated it with the golden pheasant, and in the Bible, it is associated with the ibis or peacock.

Its cult originated in Egypt, and important meanings were attributed to it, making it a bird of good omen and great spiritual significance.

The Phoenix was associated with the sun god Ra, becoming its emblem, so much so that the Bennu (the initial name that later changed to Phoenix) became the hieroglyph representing the deity of the sun.

Contrary to what the name may suggest, according to legends, the Phoenix is only male.

Famous for being the bird that rises from its own ashes, it became a symbol of the resurrection of Christ.

The legend tells that when the Phoenix felt near death, it gathered aromatic herbs such as sandalwood, myrtle, myrrh, cinnamon, and built a large nest shaped like an egg and let itself die there, burned by its own flames.

From its ashes, an egg was born that the sun made hatch and open in three days, giving life to a new Phoenix that flew away immediately.

The temple of Heliopolis, where the bird BenBen, the Arabian Phoenix, was venerated, symbolized the Great Primordial Hill, emerging from the waters of the Flood, but also the true place of origin of the resurrected bird, where the secret ritual of its resurrection took place.

The pelican symbolizes Christ who gives his body as food and his blood as a drink during the Last Supper.

The reason is linked to an ancient legend according to which this bird fed its young with its own flesh and blood.

In fact, it is curious how this seabird holds the caught food in a pouch under its beak and, upon reaching the nest, feeds the chicks by bending its beak towards the chest to extract the little fish.

The ancients, erroneously, thought that the animal tore its flesh to extract the blood with which to feed the hungry pelican chicks.

For this reason, during the Middle Ages, the pelican became a symbol of the selflessness with which children are loved and made an allegory of the supreme sacrifice of Christ, crucified on the Cross and pierced in the side from which blood and water flowed, a source of life for the salvation of men.

This engraving is not the only one in Gubbio; another exists, engraved on an old medieval portal on Via Nicola Vantaggi, in the heart of the San Martino district.

This engraving seems decidedly to represent more a pelican than an Arabian Phoenix.

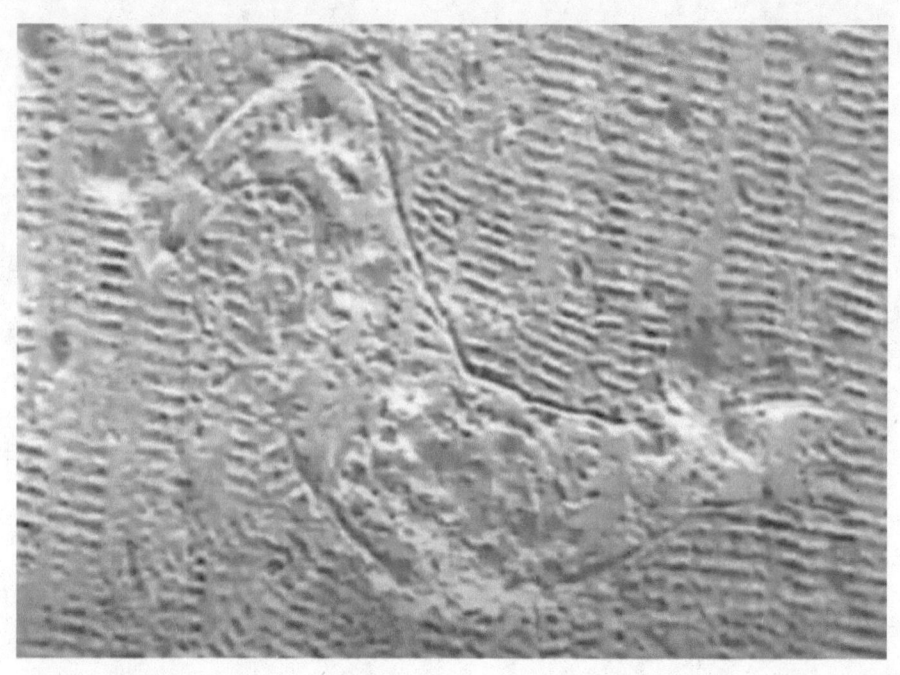

The only thing that both engravings have in common is the direction of the beak, of the head, in both cases oriented to the left.

In the engraving on the Cathedral of Gubbio, the beak is effectively directed towards the

East, as this is the orientation of the Church. In that present on the portal of Via Nicola Vantaggi, the orientation is South/West.

However, this does not detract from the fact that it could be a symbolic orientation not related to the actual position of the building.

For those who would like to see other symbologies present in the city of Gubbio, I recommend the video on YouTube: "Templar Symbolism in Gubbio.

The Tau of the Hospitallers engraved together with the Bennu bird."

If you have any questions,

my email is s1er050@yahoo.com

www.ingramcontent.com/pod-product-compliance
Lightning Source LLC
Chambersburg PA
CBHW010722110626
46523CB00046B/718